FROM **MILK** TO **CHEESE**

by Shannon Zemlicka

Lerner Publications Company / Minneapolis

Lerner Publications Company
A division of Lerner Publishing Group
241 First Avenue North
Minneapolis, MN 55401 U.S.A.

Website address: www.lernerbooks.com

Library of Congress Cataloging-in-Publication Data

Zemlicka, Shannon.
   From milk to cheese / by Shannon Zemlicka.
      p.   cm. — (Start to finish)
   Includes index.
   Summary: Briefly introduces the process by which cheese is made from milk.
   ISBN: 0–8225–1387–0 (lib. bdg. : alk. paper)
   1. Cheese—Juvenile literature. 2. Milk—Juvenile literature. [1. Cheese. 2. Milk.] I. Title. II. Start to finish (Minneapolis, Minn.)
   SF271.Z46 2004
   641.3'73—dc21   2003005607

Manufactured in the United States of America
1 2 3 4 5 6 – DP – 09 08 07 06 05 04

The photographs in this book appear courtesy of:
© Todd Strand/Independent Picture Service, cover, pp. 1 (bottom), 3; © Karlene Schwartz, p.1 (top); © Bill Tarpenning/USDA, p. 5; Milch & Markt Informationsbüro, p. 7; © Owen Franken/CORBIS, pp. 9, 19; © Inga Spence/Visuals Unlimited, pp. 11, 15, 21; © Edmond Van Hoorick/SuperStock, p. 13; © Jacqui Hurst/CORBIS, p. 17;© Royalty-Free/ CORBIS, p. 23

Table of Contents

Cheese is fun to eat!

How is it made?

A farmer milks cows.

Cheese starts as milk. Milk comes from cows, goats, or other animals. A farmer uses a machine to milk cows. The milk flows into a big tank.

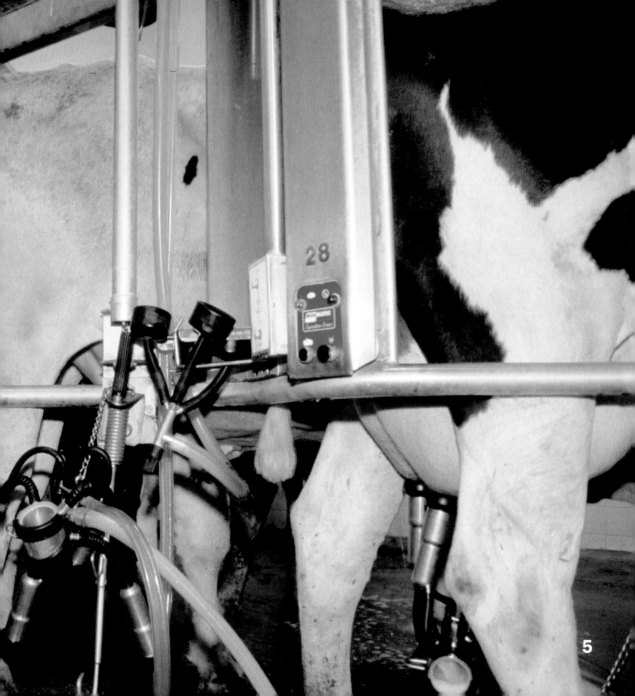

The milk is heated.

Trucks take the milk to a cheese **factory**. A factory is a place where things are made. Pipes carry the milk into big steel pots called **vats**. The milk is heated. Heating milk makes it safe to use in cheese.

The milk turns sour.

A worker adds a liquid that makes the milk turn sour. Another liquid called **rennet** makes the milk thick and chunky. Thick, sour chunks of milk are called **curd**.

Workers chop the curd.

Workers chop the curd into many pieces. The pieces are stirred and heated. A liquid called **whey** flows out of the pieces of curd.

The curd is drained.

A worker wraps a thin cloth with small holes around the curd and whey. The cloth is lifted out of the vat. The whey drips through the holes in the cloth. The curd is kept to be made into cheese.

Workers salt the curd.

A worker adds salt to the curd.
Salt helps cheese taste good.
Salt also helps cheese last a long
time without spoiling.

Machines press the curd.

The curd is pressed into a container. A lid presses down on the curd and squeezes it. The curd hardens into cheese. The cheese is the same shape as the container.

17

The cheese sits.

The cheese is put in a room to sit for days or weeks. Sitting helps cheese taste even better.

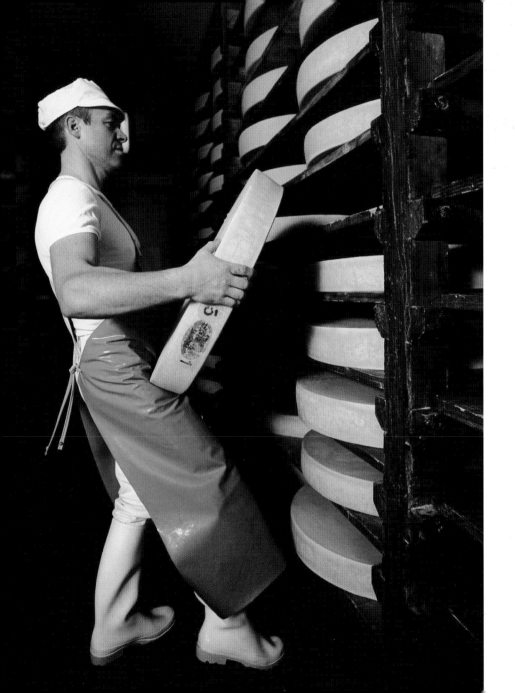

The cheese is wrapped.

Some cheese is sold in one big piece. Other cheese is sold in slices. Both kinds of cheese are wrapped in plastic. Trucks take the cheese to stores to be sold.

Eat the cheese!

Cheese tastes great in a sandwich or on a cracker. It tastes great with macaroni, too! What is your favorite way to eat cheese?

Glossary

curd (KURD): thick, sour chunks of milk

factory (FAK-tur-ee): a place where things are made

rennet (REHN-eht): a liquid that makes milk thick and chunky

vats (VATS): big steel pots

whey (WAY): a watery liquid found in thick, sour milk

Index